INTERNATIONAL CENTRE FOR MECHANICAL SCIENCES

COURSES AND LECTURES - No. 112

CHAO CHEN WANG
UNIVERSITY OF HOUSTON - TEXAS

FIELD EQUATIONS FOR THERMOELASTIC BODIES WITH UNIFORM SYMMETRY

ACCELERATION WAVES IN ISOTROPIC THERMOELASTIC BODIES

COURSE HELD AT THE DEPARTMENT
OF MECHANICS OF SOLIDS
JUNE - JULY 1971

Springer-Verlag Wien GmbH 1971

Originally published by CISM, Udine in 1972.
ISBN 978-3-662-37540-2 ISBN 978-3-662-38315-5 (eBook)
DOI 10.1007/978-3-662-38315-5

FIELD EQUATIONS FOR THERMOELASTIC BODIES WITH UNIFORM SYMMETRY

P R E F A C E

The structure of a body of uniform symmetry and, in particular, that of an isotropic body are explained in the preceding paper. In this paper we consider the theory of wave propagations in an isotropic thermoelastic body. Propagation conditions and amplitude equations for acceleration waves are derived.

Udine, July 1971

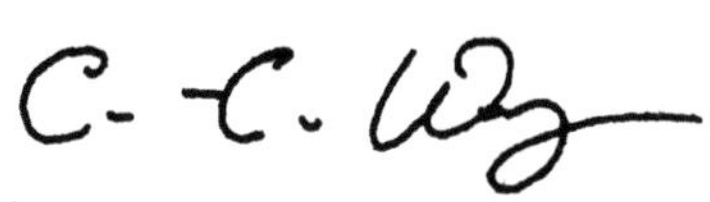

1. Introduction

In a purely mechanical theory of inhomogeneous bodies formulated recently by Noll [1] and Wang [2] two basic assumptions are used: material uniformity and smoothness. In this paper, I generalize that theory in two respects: First, I replace the assumption of material uniformity by a similar but somewhat weaker assumption of symmetry uniformity. That is, I require the points of a body to have the same material symmetry rather than the same material response. Clearly, material uniformity implies symmetry uniformity, but the converse is not true in general. Second, I consider thermodynamical state variables and state functions in this paper, and I derive the thermodynamical field equations for inhomogeneous bodies.

For bodies with uniform symmetry, we define the concept of inhomogeneity in almost the same way as that of materially uniform bodies. Specifically, a configuration of a body is said to be homogeneous in symmetry (vs. homogeneous in response for a materially uniform body) if the symmetry group of the points of the body relative to the reference configuration are independent of the position in that configuration. If such a configuration exists, the body is called (symmetry) homogeneous; otherwise, it is called (symmetry) inhomogeneous.

The concept of smoothness for bodies with uniform symmetry is also defined in almost the same way as that of materially uniform bodies. A smooth field of local reference configurations is called a reference chart for symmetry (vs. reference chart for response for a materially uniform body) if the symmetry groups relative to the local reference configuration are the same group; that group is then called the symmetry group of the reference chart. A body with uniform symmetry is said to be smooth if every point of the body can be covered by a reference chart for symmetry. Of course, in order to use field equations, we assume also that the response functions are smooth with respect to reference charts.

Geometric structures of smooth bodies with uniform symmetry are also similar to those of smooth materially uniform bodies. We can characterize the geometric structure of a smooth body with uniform symmetry by a symmetry bundle (vs. a material bundle for a materially uniform body) and a bundle of symmetry frames (vs. a bundle of reference frames for a materially uniform body). The atlases of these bundles, of course, are formed by reference charts for symmetry, and hence are called symmetry atlases (vs. material atlases for a materially uniform body).

For bodies with uniform symmetry we introduce symmetry connections (vs. material connections) by the condition that their parallel transports preserve the symmetry bundles. Departure of these connections from the Euclidean connection, then, characterizes the local inhomogeneities of the bodies.

A thermodynamical theory can be developed for

smooth bodies with uniform symmetry that are made up of general non-linear materials with long range memory effects. For simplicity, however, we consider in this paper only such bodies that are made up of thermoelastic materials.

2. Constitutive Relations of a Thermoelastic Material

We assume the usual constitutive relations for a thermoelastic material. For state functions we choose the stress tensor $\underset{\sim}{T}$, the Helmholtz free energy u , the entropy η , and the heat flux $\underset{\sim}{f}$, and for state variables we choose the deformation gradient $\underset{\sim}{F}$ relative to some local reference configuration $\underset{\sim}{\nu}$, the absolute temperature ϑ, and the temperature gradient $\underset{\sim}{g}$. Thus the constitutive relations are expressed by

$$\left(\underset{\sim}{T}, u, \eta, \underset{\sim}{f}\right) = \Phi_{\underset{\sim}{\nu}}\left(F, \vartheta, \underset{\sim}{g}\right), \tag{2.1}$$

where the components of $\Phi_{\underset{\sim}{\nu}}$ are the response functions of the material.

The response functions of a thermoelastic material are subject to restrictions due to the principle of material frame-indifference and the principle of universal dissipation. The former principle requires that

$$\left(\underset{\sim}{Q}\underset{\sim}{T}\underset{\sim}{Q}^{T}, u, \eta, \underset{\sim}{Q}\underset{\sim}{f}\right) = \Phi_{\underset{\sim}{\nu}}\left(\underset{\sim}{Q}\underset{\sim}{F}, \vartheta, \underset{\sim}{Q}\underset{\sim}{g}\right) \tag{2.2}$$

for all orthogonal tensors $\underset{\sim}{Q}$, while the latter principle requires that

(2.3)
$$(\underset{\sim}{T}, u, \eta) = \left(\rho \underset{\sim}{F}\left(\frac{\partial u}{\partial \underset{\sim}{F}}\right)^T, u, -\frac{\partial u}{\partial \vartheta}\right) = \Psi_{\underset{\sim}{\nu}}(\underset{\sim}{F}, \vartheta),$$
$$f_{\underset{\sim}{\nu}}(\underset{\sim}{F}, \vartheta, \underset{\sim}{g}) \cdot \underset{\sim}{g} \geq 0 .$$

Symmetry of a thermoelastic material may be characterized by the symmetry groups of the response functions or simply by the intersection of those groups. That intersection is a subgroup of the unimodular group consisting in tensors $\underset{\sim}{P}$ satisfying the symmetry condition:

(2.4)
$$\Phi_{\nu}(\underset{\sim}{F}\underset{\sim}{P}, \vartheta, \underset{\sim}{g}) = \Phi_{\underset{\sim}{\nu}}(\underset{\sim}{F}, \vartheta, \underset{\sim}{g})$$

for all $(\underset{\sim}{F}, \vartheta, \underset{\sim}{g})$. We call this subgroup the symmetry group of the thermoelastic material relative to $\underset{\sim}{\nu}$, and we denote it by $g_{\underset{\sim}{\nu}}$

3. Geometric Structure of a Smooth Thermoelastic Body with Uniform Symmetry

For a smooth thermoelastic body $\mathcal{B}$ with uniform symmetry, there exists a global field of local reference configurations $\underset{\sim}{\nu}$ relative to which the symmetry groups of the points of $\mathcal{B}$ are the same group $g_{\underset{\sim}{\nu}}$. The response functions of the points of $\mathcal{B}$ relative to $\underset{\sim}{\nu}$, however, need not be the same function, since we do not require $\mathcal{B}$ to be materially uniform. Thus the response functions form a field $\Phi_{\underset{\sim}{\nu}}(\underset{\sim}{F}, \vartheta, \underset{\sim}{g}, X)$

where X is a typical point in $\mathcal{B}$.

Generally, the global field $\underset{\sim}{\nu}$ that gives rise to the uniform symmetry group $g_{\underset{\sim}{\nu}}$ need not be a smooth field. The assumption that $\mathcal{B}$ is smooth, however, requires that, locally, there exists smooth fields $\underset{\sim}{\mu}$ differing from the global field $\underset{\sim}{\nu}$ by deformation gradients belonging to the symmetry group $g_{\underset{\sim}{\nu}}$. Let $\underset{\sim}{\mu}$ be such a smooth field defined on a subbody U of $\mathcal{B}$. Then the pair $(U, \underset{\sim}{\mu})$ is called a <u>reference chart for symmetry</u> covering the subbody U. Since from (2.4) the response functions relative to $\underset{\sim}{\mu}$ are still the functions $\Phi_{\underset{\sim}{\nu}}$, the symmetry groups relative to $\underset{\sim}{\mu}$ are still the group g_{ν}. We denote the collection of all such reference charts by

$$\mathcal{A} = \left\{ (U_\alpha, \underset{\sim}{\mu}_\alpha), \alpha \in A \right\}, \tag{3.1}$$

where A is an index set. We call $\mathcal{A}$ <u>a reference atlas</u> for $\mathcal{B}$. Since $\Phi_{\underset{\sim}{\nu}}$ is the field of response functions and $g_{\underset{\sim}{\nu}}$ is their symmetry group for all reference charts in $\mathcal{A}$, we now rewrite them as $\Phi_{\mathcal{A}}$ and $g_{\mathcal{A}}$ respectively, and we call them the <u>field of response functions</u> and the <u>symmetry group</u> relative to $\mathcal{A}$.

From the definition of a reference chart in $\mathcal{A}$ we see that the deformation gradient $\underset{\sim}{G}_{\alpha\beta}$ from $(U_\alpha, \underset{\sim}{\mu}_\alpha)$ to $(U_\beta, \underset{\sim}{\mu}_\beta)$ is a smooth field on $U_\alpha \cap U_\beta$ with values in $g_{\mathcal{A}}$. Such fields are called the <u>co-ordinate trasformations</u> of $\mathcal{A}$

Clearly, they satisfy the identities

$$
\begin{aligned}
\underset{\sim}{G}_{\alpha\alpha} &= I \quad \text{on } U_\alpha, \\
\underset{\sim}{G}_{\alpha\beta} &= \underset{\sim}{G}^{-1}_{\beta\alpha} \quad \text{on } U_\alpha \cap U_\beta, \qquad (3.2)\\
\underset{\sim}{G}_{\alpha\beta}\underset{\sim}{G}_{\beta\gamma} &= \underset{\sim}{G}_{\alpha\gamma} \quad \text{on } U_\alpha \cap U_\beta \cap U_\gamma .
\end{aligned}
$$

In view of these identities, we see that $\mathcal{A}$ can be regarded as the bundle atlas of a subbundle $T(\mathcal{B},\mathcal{A})$ of the tangent bundle $T(\mathcal{B})$ of $\mathcal{B}$, the symmetry group of the subbundle $T(\mathcal{B},\mathcal{A})$ being simply the symmetry group $g_{\mathcal{A}}$. We call $T(\mathcal{B},\mathcal{A})$ the <u>symmetry bundle</u> of $\mathcal{B}$ relative to $\mathcal{A}$, and we call the associated principal bundle $E(\mathcal{B},\mathcal{A})$ of $T(\mathcal{B},\mathcal{A})$ the <u>bundle of symmetry frames</u> of $\mathcal{B}$ relative to $\mathcal{A}$. As in the theory of materially uniform bodies, these bundles characterize completely the material geometric structure of $\mathcal{B}$.

An affine connection on $T(\mathcal{B})$ reducible to a connection on $T(\mathcal{B},\mathcal{A})$ is called a <u>symmetry connection</u> of $\mathcal{B}$ relative to $\mathcal{A}$. The theory of symmetry connections is essentially the same as that of material connections for materially uniform bodies. Hence we shall not repeat the theory here.

4. Field Equations

For definiteness, we choose a fixed rectangular Cartesian co-ordinate system in space and a particular reference atlas $\mathcal{A}$ for $\mathcal{B}$, so that we can suppress $\mathcal{A}$ from the notations. Relative to the fixed co-ordinate system, we define

$$T^{ij}{}_{kl} \equiv T^{ij}{}_{kl}(\underset{\sim}{F}, \vartheta, X) \equiv \frac{\partial T^{ij}}{\partial F^{kl}} . \tag{4.1}$$

From the condition (2.4), we then have

$$\begin{aligned} &T^{ij}{}_{kl}(\underset{\sim}{F}, \vartheta, X)\, F^{km} K_m{}^{l} = 0 \\ &T^{ij}{}_{kl}(\underset{\sim}{F}, \vartheta, X) = T^{ij}{}_{km}(\underset{\sim}{F}\underset{\sim}{P}, \vartheta, X)\, P_l{}^{m} \end{aligned} \tag{4.2}$$

for all $\underset{\sim}{K}$ belonging to the Lie algebra(*) of $\mathcal{g}$ and for all $\underset{\sim}{P}$ belonging to $\mathcal{g}$.

Now let $\underset{\sim}{\chi}$ be a configuration of $\mathcal{B}$ with co-ordinate functions (x^i), viz

$$\underset{\sim}{x} = \underset{\sim}{\chi}(X) = \left(x^1(X), x^2(X), x^3(X)\right). \tag{4.3}$$

Then the stress tensor at $\underset{\sim}{x}$ can be determined in the following way: Choose any reference chart $(U_\alpha, \underset{\sim}{\mu}_\alpha) \in \mathcal{A}$ covering X and let $\underset{\sim}{F}$ be the deformation gradient from $\underset{\sim}{\mu}_\alpha$ to $\underset{\sim}{\chi}$ at X . Then $\underset{\sim}{T}|_{\underset{\sim}{x}}$ is given by the constitutive relation (2.1).

(*) Since the response functions are assumed to be smooth, the symmetry group $\mathcal{g}$ is a Lie subgroup of the unimodular group.

Substituting $\underset{\sim}{T}$ into the equation of balance of linear momentum

(4.4) $$\operatorname{div} \underset{\sim}{T} + \rho \underset{\sim}{b} = \rho \underset{\sim}{a} ,$$

where $\underset{\sim}{b}$ and $\underset{\sim}{a}$ are the body force and the acceleration, we get

(4.5) $$T^{ij}{}_{kl} \frac{\partial F^{kl}}{\partial x^j} + T^{ij}_{\vartheta} g_j + T^i + \rho b^i = \rho a^i$$

where

(4.6) $$T^{ij}_{\vartheta} \equiv T^{ij}_{\vartheta}(\underset{\sim}{F}, \vartheta, \underset{\sim}{x}) \equiv \frac{\partial T^{ij}}{\partial \vartheta},$$
$$T^i \equiv T^i(\underset{\sim}{F}, \vartheta, \underset{\sim}{x}) \equiv \frac{\partial T^{ij}}{\partial x^j},$$

Here we have replaced the argument X by the co-ordinates (x^i). Like $T^{ij}{}_{kl}$, the functions T^{ij}_{ϑ} and T^i obey the conditions

(4.7) $$T^{ij}_{\vartheta}(\underset{\sim}{F}, \vartheta, \underset{\sim}{x}) = T^{ij}_{\vartheta}(\underset{\sim}{F}\underset{\sim}{P}, \vartheta, \underset{\sim}{x})$$
$$T^i(\underset{\sim}{F}, \vartheta, \underset{\sim}{x}) = T^i(\underset{\sim}{F}\underset{\sim}{P}, \vartheta, \underset{\sim}{x})$$

for all $P \in g$. Since μ_α and F are defined on the subbody U_α only, so the gradient $\partial F^{kl}/\partial x^j$. Hence (4.5) is a local equation, valid for the subbody U_α only.

To replace the local equation (4.5) by a global one, we use a technique developed in the theory of materially uniform bodies. We choose a symmetry connection relative to $\mathfrak{A}$. Let Γ^i_{jk} be a connection symbol of the symmetry connection relative to the co-ordinate system (x^i) Then the fields K_m with component matrices

$$\left[(K_m)^i{}_k\right] = \left[F^{-1\,i}{}_j\left(\frac{\partial F^j{}_k}{\partial x^m} + \Gamma^j_{\ell m}F^\ell{}_k\right)\right] \qquad m = 1,2,3, \qquad (4.8)$$

have values in the Lie algebra of g . Hence from $(4.2)_1$, we have

$$T^{ij}{}_{k\ell}\left(\frac{\partial F^{k\ell}}{\partial x^m} + \Gamma^k_{nm}F^{n\ell}\right) = 0 \ . \qquad (4.9)$$

Substituting (4.9) into (4.5), we obtain

$$-T^{ij}{}_{k\ell}F^{n\ell}\Gamma^k_{nj} + T^{ij}_{\vartheta}g_j + T^i + \rho b^i = \rho a^i . \qquad (4.10)$$

Now we define

$$\bar{T}^{ij}{}_k{}^n \equiv T^{ij}{}_{k\ell}(\underset{\sim}{F}, \vartheta, \underset{\sim}{x})\, F^{n\ell} . \qquad (4.11)$$

Then from $(4.2)_2$ we have

$$\bar{T}^{ij}{}_k{}^n(\underset{\sim}{F}, \vartheta, \underset{\sim}{x}) = \bar{T}^{ij}{}_k{}^n(\underset{\sim}{F}\underset{\sim}{P}, \vartheta, \underset{\sim}{x}) \qquad (4.12)$$

for all $\underset{\sim}{P} \in g$.

Now from (4.7), (4.12), and the condition that the co-ordinate transformations of $\mathfrak{A}$ are fields in g , we see that the fields $\bar{T}^{ij}{}_k{}^n$, T^{ij}_{ϑ} ,and T^i are independent of the choice of the reference chart. Thus a global field equation is

$$(4.13)\qquad -\bar{T}^{ij}{}_{k}\,{}^{n}\Gamma^{k}_{nj} + T^{ij}_{\vartheta}\, g_j + T^{i} + \rho b^{i} = \rho a^{i} .$$

This global field equation can be rewritten in several equivalent forms. First, if we introduce a fixed reference configuration corresponding to co-ordinate system $\underset{\sim}{K} = (X^A)$, then the co-ordinates (x^i) are given by the deformations functions

$$(4.14)\qquad x^{i} = x^{i}(X^{A}, t).$$

In this case (4.13) can be rewritten in the form

$$(4.15)\qquad \begin{aligned} &\bar{T}^{ij}{}_{k}{}^{A}\left(\frac{\partial^2 x^k}{\partial X^A \partial X^B} - \Gamma^{C}_{AB}\frac{\partial x^k}{\partial X^C}\right)\frac{\partial X^B}{\partial x^j} + T^{ij}_{\vartheta}\, g_j \\ &+ T^{i} + \rho b^{i} = \rho a^{\ell}, \end{aligned}$$

where Γ^{A}_{BC} is a connection symbol of the symmetry connection relative to the co-ordinate system (X^A) and where $\bar{T}^{ij}{}_{k}{}^{A}$ is defined by

$$(4.16)\qquad \bar{T}^{ij}{}_{k}{}^{A} \equiv T^{ij}{}_{k\ell}\left(\underset{\sim}{F}, \vartheta, \underset{\sim}{x}\right) F^{Ak},$$

$\bar{F}^{Ak}$ being a component of the deformation gradient from $\underset{\sim}{\mu}_{\alpha}$ to $\underset{\sim}{K}$.

Next, if we introduce the Piola-Kirchhoff stress tensor $\underset{\sim}{R}$ relative to $\mathcal{Q}$ by

$$\underset{\sim}{R} \equiv \underset{\sim}{R}(\underset{\sim}{F}, \vartheta, \underset{\sim}{X}) = (\det \underset{\sim}{F})\, \underset{\sim}{T}(\underset{\sim}{F}, \vartheta, \underset{\sim}{X})(\underset{\sim}{F}^{-1})^{T}, \tag{4.17}$$

then the field equation (4.13) can be rewritten in the form

$$\begin{aligned} &\tilde{R}^{iA}{}_{k}{}^{B}\left(\frac{\partial^2 x^k}{\partial X^A \partial X^B} - \Gamma^{C}_{AB}\frac{\partial x^k}{\partial X^C}\right) - \tilde{R}^{iA}\, C^{B}_{AB} + \\ &+ \tilde{R}^{iA}_{\vartheta}\, g_A + \tilde{R}^{i} + \rho_{\underset{\sim}{k}} b^i = \rho_{\underset{\sim}{k}} a^i, \end{aligned} \tag{4.18}$$

where

$$\begin{aligned} \tilde{R}^{iA}{}_{k}{}^{B} &\equiv \frac{1}{\det \underset{\sim}{F}} R^{ir}{}_{k\ell}(\underset{\sim}{F}, \vartheta, \underset{\sim}{X}) \tilde{F}^{A}{}_{r} \tilde{F}^{B\ell}, \\ R^{ir}{}_{k\ell} &\equiv R^{ir}{}_{k\ell}(\underset{\sim}{F}, \vartheta, \underset{\sim}{X}) = \frac{\partial R^{ir}}{\partial F^{k\ell}}, \\ \tilde{R}^{iA} &\equiv \frac{1}{\det \underset{\sim}{F}} R^{ir}(\underset{\sim}{F}, \vartheta, \underset{\sim}{X}) \tilde{F}^{A}{}_{r}, \\ \tilde{R}^{iA}_{\vartheta} &\equiv \frac{\partial \tilde{R}^{iA}}{\partial \vartheta}, \\ \tilde{R}^{i} &\equiv \frac{\partial \tilde{R}^{iA}}{\partial X^A}, \\ C^{A}_{BC} &\equiv \Gamma^{A}_{BC} - \Gamma^{A}_{CB}. \end{aligned} \tag{4.19}$$

The same technique can be used to derive the field equation for energy balance. We define

$$(4.20) \qquad f^{i}_{jk} \equiv f^{i}_{jk}(\underset{\sim}{F}, \vartheta, \underset{\sim}{g}, X) \equiv \frac{\partial f^{i}}{\partial F^{jk}} .$$

Then the following two conditions hold:

$$(4.21) \qquad \begin{aligned} & f^{i}_{jk}(\underset{\sim}{F}, \vartheta, \underset{\sim}{g}, X) F^{jm} K_{m}^{\;k} = 0 \\ & f^{i}_{jm}(\underset{\sim}{F}, \vartheta, \underset{\sim}{g}, X) = f^{i}_{jk}(\underset{\sim}{F}\underset{\sim}{P}, \vartheta, g, X) P_{m}^{\;k} \end{aligned}$$

for all $\underset{\sim}{K}$ belonging to the Lie algebra of g and for all $\underset{\sim}{P}$ belonging to g.

In a configuration $\underset{\sim}{\chi}$ corresponding to co-ordinate system (x^{i}), the entropy and the heat flux at $\underset{\sim}{x}$ can be determined by the constitutive relation (2.1) with deformation gradient $\underset{\sim}{F}$ taken relative to any reference chart $(U_{\alpha}, \mu_{\alpha})$ covering X. Substituting η and $\underset{\sim}{f}$ into the equation of balance of energy

$$(4.22) \qquad \operatorname{div} \underset{\sim}{f} + \rho r = \rho \vartheta \dot{\eta} ,$$

where r is the energy supply, we get

$$(4.23) \qquad g^{i}_{jk} \frac{\partial F^{jk}}{\partial x^{i}} + f^{i}_{\vartheta} g_{i} + f^{ij}_{\underset{\sim}{g}} \frac{\partial g_{j}}{\partial x^{i}} + f + \rho r =$$

$$= \rho\vartheta\left(\eta_{jk}\dot{F}^{jk} + \eta_{\vartheta}\dot{\vartheta}\right)$$

where

$$f_{\vartheta}^{i} \equiv f_{\vartheta}^{i}(\underset{\sim}{F}, \vartheta, \underset{\sim}{g}, \underset{\sim}{x}) \equiv \frac{\partial f^{i}}{\partial \vartheta},$$

$$f_{g}^{ij} \equiv f_{g}^{ij}(\underset{\sim}{F}, \vartheta, \underset{\sim}{g}, x) \equiv \frac{\partial g^{i}}{\partial g_{j}},$$

$$f \equiv f(\underset{\sim}{F}, \vartheta, \underset{\sim}{g}, \underset{\sim}{x}) \equiv \frac{\partial g^{i}}{\partial x^{i}}, \qquad (4.24)$$

$$\eta_{jk} \equiv \eta_{jk}(\underset{\sim}{F}, \vartheta, \underset{\sim}{x}) \equiv \frac{\partial \eta}{\partial F^{jk}},$$

$$\eta_{\vartheta} \equiv \eta_{\vartheta}(\underset{\sim}{F}, \vartheta, \underset{\sim}{x}) \equiv \frac{\partial \eta}{\partial \vartheta}.$$

As before the field equation (4.23) is valid for the subbody U_{α} only.

By the same argument as before, if we introduce a symmetry connection relative to $\mathcal{A}$, then we can rewrite (4.23) as

$$-\bar{f}^{i}{}_{j}{}^{n}\Gamma^{j}_{ni} + f^{i}_{\vartheta} g_i + f^{ij}_{g}\frac{\partial g_j}{\partial x^i} + f + \rho r = \tag{4.25}$$

$$= \rho\vartheta\left(\eta_{jk}\dot{F}^{jk} + \eta_{\vartheta}\,\vartheta\right),$$

where locally we have

$$\bar{f}^{i}{}_{j}{}^{n} = f^{i}{}_{jk}(\underset{\sim}{F}, \vartheta, \underset{\sim}{g}, \underset{\sim}{x})\; F^{nk} \tag{4.26}$$

but $\bar{f}^{i}{}_{j}{}^{n}$ is independent of the choice of reference chart. If we now introduce a reference configuration $\underset{\sim}{K}$ corresponding to co-ordinate system (X^A), then we can rewrite (4.26) as

$$\tilde{f}^{i}{}_{k}{}^{A}\left(\frac{\partial^2 x^k}{\partial X^A \partial X^B} - \Gamma^{C}_{AB}\frac{\partial x^k}{\partial X^C}\right)\frac{\partial X^B}{\partial x^i} + f^{i}_{\vartheta} g_i + f^{ij}_{g}\frac{\partial g_j}{\partial x^i} \tag{4.27}$$

$$+ f + \rho r = \rho\vartheta\left(\tilde{\eta}_i{}^{A}\frac{\partial x^l}{\partial X^A}\frac{\partial v^i}{\partial x^l} + \eta_{\vartheta}\dot{\vartheta}\right),$$

where locally we have

$$\tilde{f}^{i}{}_{k}{}^{A} \equiv f^{j}{}_{kl}(\underset{\sim}{F}, \vartheta, \underset{\sim}{g}, \underset{\sim}{x})\;\tilde{F}^{Ak}\,,$$

(4.28)

$$\tilde{\eta}_j{}^{A} \equiv \eta_{jk}(\underset{\sim}{F}, \vartheta, \underset{\sim}{x})\;\tilde{F}^{Ak}\,.$$

The equation (4.27) now has a global form.

Finally, we can introduce the heat flux function

$\underset{\sim}{p}$ relative to $\mathcal{A}$ by

$$\underset{\sim}{p} \equiv \underset{\sim}{p}(\underset{\sim}{F}, \vartheta, \underset{\sim}{g}, X) \equiv (\det \underset{\sim}{F})\underset{\sim}{F}^{-1} f(\underset{\sim}{F}, \vartheta, \underset{\sim}{g}, X). \tag{4.29}$$

Then the field equation (4.27) can be rewritten in the form

$$\begin{aligned} &\tilde{p}_k^{AB}\left(\frac{\partial^2 x^k}{\partial X^A \partial X^B} - \Gamma^C_{BA}\frac{\partial x^k}{\partial X^C}\right) - \tilde{p}^A C^B_{AB} + \tilde{p}^A_\vartheta g_A + \\ &+ \tilde{p}_{\underset{\sim}{g}}^{AB}\frac{\partial g_B}{\partial X^A} + \tilde{p} + \rho_{\underset{\sim}{k}} r = \rho_k \vartheta\left(\tilde{\eta}_i^A \frac{\partial x^\ell}{\partial X^A}\frac{\partial v^i}{\partial x^\ell} + \eta_\vartheta \dot{\vartheta}\right) \end{aligned} \tag{4.30}$$

where

$$\begin{aligned} \tilde{p}_k^{AB} &\equiv \frac{1}{\det \underset{\sim}{F}} p^r_{k\ell}(\underset{\sim}{F}, \vartheta, \underset{\sim}{g}, \underset{\sim}{X}) \tilde{F}^A{}_r F^{B\ell}, \\ p^r_{k\ell} &\equiv p^r_{k\ell}(\underset{\sim}{F}, \vartheta, \underset{\sim}{g}, \underset{\sim}{X}) \equiv \frac{\partial p^r}{\partial F^{k\ell}}, \\ \tilde{p}^A &\equiv \frac{1}{\det \underset{\sim}{F}} p^i(\underset{\sim}{F}, \vartheta, g, \underset{\sim}{X}) \tilde{F}^A{}_i, \\ \tilde{p}_{\underset{\sim}{g}}^{AB} &= \frac{\partial \tilde{p}^A}{\partial g_B}, \\ \tilde{p}^A_\vartheta &\equiv \frac{\partial \tilde{p}^A}{\partial \vartheta}, \\ \tilde{p} &\equiv \frac{\partial \tilde{p}^A}{\partial X^A}. \end{aligned} \tag{4.31}$$

References

[1] W. Noll, Arch. Rational Mech. Anal. 27, 1-32 (1967). Reprinted in Continuum Theory of Inhomogeneities in Simple Bodies, Springer-Verlag (1968)

[2] C.-C. Wang, Arch. Rational Mech. Anal. 27, 33-94 (1967)

Reprinted in Continuum Theory of Inhomogeneities in Simple Bodies, Springer-Verlag (1968)

[3] C.-C. Wang, "Thermodynamics of Inhomogeneous Bodies" Proc. Int'l Conf. "Fundamental Aspects of Dislocations," National Bureau of Standards, April 1970.

ACCELERATION WAVES IN ISOTROPIC THERMOELASTIC BODIES

P R E F A C E

A body with uniform symmetry is defined by the condition that the symmetry groups of the points of the body are isomorphic. For example, if the points of a body are all isotropic points, then the body is a body with uniform symmetry, called an *<u>isotropic body</u>*. *In this paper, we derive the field equations for the balance of linear momentum and the balance of energy for bodies with uniform symmetry.*

Udine, July 1971

C. C. Wang

1. Introduction

This paper is based on a paper of Bowen & Wang [1] which will be published in the Archive for Rational Mechanics and Analysis shortly.

In the preceding paper [0] I have developed a theory of inhomogeneous thermoelastic bodies in general. In this paper I shall consider wave propagations in a special class of such bodies: isotropic bodies. This class of bodies is characterized by the condition that there exists reference atlases relative to which the symmetry groups are the orthogonal group. Such atlases are then called undistorted atlases.

The importance of undistorted atlases is due to the fact that relative to such atlases the response functions are isotropic functions. This fact enables us to derive some identities for the gradients of the response functions, and the identities then enable us to derive the propagation condition and the equation of growth for the amplitudes of the acceleration waves.

A thermoelastic material is called a definite conductor if the gradient of the heat flux $\underset{\sim}{f}$ with respect to the temperature gradient $\underset{\sim}{g}$ is positive-definite viz,

$$k(\underset{\sim}{e}) \equiv \underset{\sim}{e} \cdot \frac{\partial \underset{\sim}{f}}{\partial \underset{\sim}{g}} \underset{\sim}{e} > 0 \tag{1.1}$$

for all vectors $\underset{\sim}{e} \neq \underset{\sim}{0}$. This condition generalizes the condition that the heat conduction tensors in the classical Fourier's law of heat conduction are positive-definite. Coleman & Gurtin have shown that acceleration waves in a definite conductor are homothermal, which means that the temperature gradient suffers no jump discontinuity across an acceleration wave.

A thermoelastic body is called a non conductor if the heat flux vanishes identically in all configurations of the body, viz,

$$\underset{\sim}{f}(\underset{\sim}{F}, \vartheta, \underset{\sim}{g}, X) = \underset{\sim}{0}. \tag{1.2}$$

Coleman & Gurtin have shown that acceleration waves in a non conductor are homentropic, which means that the entropy gradient suffers no jump discontinuity across an acceleration wave.

Since the theories for these two kinds of waves are essentially the same, we shall consider homothermical waves only in this paper.

2. The Characteristic Riemannian Metric

Let $\mathcal{B}$ be an isotropic thermoelastic body and suppose that $\mathcal{A}$ is an undistorted atlas for $\mathcal{B}$. Then we can define a Riemannian metric $\underset{\sim}{G}$ on $\mathcal{B}$ by

$$\underset{\sim}{G}(\underset{\sim}{u}, \underset{\sim}{v}) = \underset{\sim}{\mu}_\alpha(\underset{\sim}{u}) \cdot \underset{\sim}{\mu}_\alpha(\underset{\sim}{v}), \tag{2.1}$$

where $\underset{\sim}{u}$ and $\underset{\sim}{v}$ are tangent vectors at any point $X \in \mathcal{B}$ and where $(U_\alpha, \underset{\sim}{\mu}_\alpha)$ is any reference chart in $\mathcal{A}$ such that U_α covers X. Since the co-ordinate transformations of an undistorted atlas are orthogonal tensor fields, the right-hand side of (2.1) is independent of the choice of the reference chart. Thus $\underset{\sim}{G}$ is well-defined.

The Riemannian metric $\underset{\sim}{G}$ characterizes the material geometric structure of $\mathcal{B}$ completely, since if $\underset{\sim}{G}$ is given, we can recover the reference charts (U_α, μ_α) and the undistorted atlas $\mathcal{A}$ by the condition (2.1). For this reason, we call $\underset{\sim}{G}$ the characteristic Riemannian metric of $\mathcal{B}$ relative to $\mathcal{A}$.

Now using the condition of material frame-indifference and the condition of symmetry, we see that the response functions of $\mathcal{B}$ relative to $\mathcal{A}$ are isotropic funtions.
That is, we have:

$$(\underset{\sim}{Q}\,\underset{\sim}{T}\,\underset{\sim}{Q}^T, u, \eta, \underset{\sim}{Q}\,\underset{\sim}{f}) = \Phi_{\mathcal{A}}(\underset{\sim}{Q}\,\underset{\sim}{F}, \vartheta, \underset{\sim}{Q}\,\underset{\sim}{g}, X) \tag{2.2}$$

for all orthogonal tensors $\underset{\sim}{Q}$. Combining this condition with the restriction due to universal dissipation we then get

$$\begin{aligned}
\underset{\sim}{T} &= \underset{\sim}{T}(\underset{\sim}{B}, \vartheta, X) = \underset{\sim}{Q}^T\,\underset{\sim}{T}(\underset{\sim}{Q}\,\underset{\sim}{B}\,\underset{\sim}{Q}^T, \vartheta, X)\,\underset{\sim}{Q}, \\
u &= u(\underset{\sim}{B}, \vartheta, X) = u(\underset{\sim}{Q}\,\underset{\sim}{B}\,\underset{\sim}{Q}^T, \vartheta, X), \\
\eta &= \eta(\underset{\sim}{B}, \vartheta, X) = \eta(\underset{\sim}{Q}\,\underset{\sim}{B}\,\underset{\sim}{Q}^T, \vartheta, X), \\
\underset{\sim}{f} &= \underset{\sim}{f}(\underset{\sim}{B}, \vartheta, \underset{\sim}{g}, X) = \underset{\sim}{Q}^T\,\underset{\sim}{f}(\underset{\sim}{Q}\,\underset{\sim}{B}\,\underset{\sim}{Q}^T, \vartheta, \underset{\sim}{Q}\,\underset{\sim}{g}, X),
\end{aligned} \tag{2.3}$$

where $\underset{\sim}{B}$ is the left Cauchy-Green tensor defined by

(2.4) $$\underset{\sim}{B} \equiv \underset{\sim}{F}\underset{\sim}{F}^T .$$

The second conditions of (2.3) mean that the response functions are isotropic functions with respect to the variables $\underset{\sim}{B}$ and $\underset{\sim}{g}$ also.

Now in a configuration $\underset{\sim}{k}$ of $\mathcal{B}$ corresponding to the co-ordinate system (X^A), the components of the tensor field $\underset{\sim}{B}$ of $\underset{\sim}{k}$ relative to any reference chart $(U_\alpha, \underset{\sim}{\mu}_\alpha) \in \mathcal{A}$ are precisely the components of the characteristic Riemannian metric $\underset{\sim}{G}$ in the co-ordinate system (X^A). This means, if

(2.5) $$x^i = x^i (X^A, t) , \quad i = 1, 2, 3$$

are the deformation functions of a configuration $\underset{\sim}{\chi}$ corresponding to co-ordinate system (x^i), then the components of the tensor field $\underset{\sim}{B}$ of $\underset{\sim}{\chi}$ in (x^i) are given by

(2.6) $$B^{ij} = G^{AB} \frac{\partial x^i}{\partial X^A} \frac{\partial x^j}{\partial X^B} .$$

To compute the values of the state functions in the configuration we must use the argument $\underset{\sim}{B}$ given by this formula.

The formula (2.6) characterizes the basic difference between a homogeneous isotropic body and an inhomogeneous one. For a homogeneous body, $\underset{\sim}{G}$ is the Euclidean metric with components δ^{AB}. Then (2.6) reduces to

$$B^{ij} = \frac{\partial x^i}{\partial X^A} \frac{\partial x^j}{\partial X_A} \tag{2.7}$$

Which is merely the component form of (2.4) in Cartesian co-ordinates. For an inhomogeneous body, $\underset{\sim}{G}$ is no longer flat, and G^{AB} cannot always reduce to δ^{AB} in any (X^A). In this case, the curvature tensor of $\underset{\sim}{G}$ can be used as a measure of the local inhomogeneitics of $\mathcal{B}$.

3. Propagation Conditions

For a homothermal acceleration wave in a configuration $\underset{\sim}{\chi}$ of $\mathcal{B}$ we assume that the tensor field $\underset{\sim}{B}$, the velocity field $\underset{\sim}{v}$, the temperature field ϑ, and the temperature gradient $\underset{\sim}{g}$ are continuous, where as derivates of these fields, in particular the acceleration field, may suffer some jump discontinuities at a moving surface $\mathcal{S} = \mathcal{S}_t$. Naturally, we call such a moving surface an <u>acceleration wave</u>.

To analyze acceleration waves, we use the theory of singular surfaces. Specifically, we have the following <u>geometric and kinematic conditions of combatibility</u>:

$$\begin{aligned} [\operatorname{grad} \underset{\sim}{v}] &= -\frac{1}{W}[\underset{\sim}{a}] \otimes n, \\ [\dot{\underset{\sim}{B}}] &= -\frac{1}{W}\left([\underset{\sim}{a}] \otimes \underset{\sim}{B}\underset{\sim}{n} + \underset{\sim}{B}\underset{\sim}{n} \otimes [\underset{\sim}{a}]\right), \\ [\operatorname{grad} \underset{\sim}{B}] &= \frac{1}{W^2}\left(\underset{\sim}{B}\underset{\sim}{n} \otimes [\underset{\sim}{a}] \otimes \underset{\sim}{n} + [\underset{\sim}{a}] \otimes \underset{\sim}{B}\underset{\sim}{n} \otimes \underset{\sim}{n}\right), \end{aligned} \tag{3.1}$$

where $\underset{\sim}{n}$ is the positive unit normal of $\mathcal{S}_t$ and where W is the intrinsic speed of the wave defined by

$$W \equiv \omega - \underset{\sim}{v} \cdot \underset{\sim}{n} , \tag{3.2}$$

ω being the usual displacement speed of the moving surface. Since we shall restrict our attention to waves propagating into a rest region with $\underset{\sim}{v} = \underset{\sim}{0}$ at $\mathcal{S}_t$, (3.2) implies that W are ω equal. Under this assumption we have also the following iterated geometric and kinematic conditions of compatibility:

$$(3.3) \quad \begin{aligned} & 2\frac{\delta \underset{\sim}{s}}{\delta t} - \frac{\underset{\sim}{s}}{\omega}\frac{\delta \omega}{\delta t} = \left[\frac{\partial^2 \underset{\sim}{v}}{\partial t^2}\right] - \omega^2 \underset{\sim}{C} \\ & \left[\operatorname{grad}(\operatorname{grad} \underset{\sim}{v})\right] = \underset{\sim}{C} \otimes \underset{\sim}{n} \otimes \underset{\sim}{n} - \frac{\partial(\underset{\sim}{s}/\omega)}{\partial y^\Gamma} \otimes \left(\underset{\sim}{a}^\Gamma \otimes \underset{\sim}{n} + \underset{\sim}{n} \otimes \underset{\sim}{a}^\Gamma\right) + (\underset{\sim}{s}/\omega) \otimes \underset{\sim}{\Omega}, \end{aligned}$$

where $\underset{\sim}{s}$ is the amplitude vector defined by

$$\underset{\sim}{s} \equiv \left[\frac{\partial \underset{\sim}{s}}{\partial t}\right] = [\underset{\sim}{a}] \tag{3.4}$$

$\underset{\sim}{C}$ denotes the vector

$$\underset{\sim}{C} = \left[\operatorname{grad}(\operatorname{grad} \underset{\sim}{v})\right](\underset{\sim}{n}, \underset{\sim}{n}), \tag{3.5}$$

$\delta/\delta t$ denotes the displacement derivate with respect to the mo ving surface, $(y^\Gamma) = (y^1, y^2)$ denotes a surface co-ordinate sys tem with natural dual basis $(\underset{\sim}{a}^1, \underset{\sim}{a}^2)$ and $\underset{\sim}{\Omega}$ denotes the second fundamental form of the surface.

As in the preceding paper [0] we have the field equation

$$H^{km}_{\ \ pf} \frac{\partial B^{pf}}{\partial x^m} + H^{km}_{\vartheta}\, g_m + H^k + \rho b^k = \rho a^k , \tag{3.6}$$

where

$$H^{km}_{\ \ pf} \equiv H^{km}_{\ \ pf}(\underset{\sim}{B}, \underset{\sim}{\vartheta}, \underset{\sim}{x}) \equiv \frac{\partial T^{km}}{\partial B^{pf}} ,$$

$$H^{km}_{\vartheta} \equiv H^{km}_{\vartheta}(\underset{\sim}{B}, \underset{\sim}{\vartheta}, \underset{\sim}{x}) \equiv \frac{\partial T^{km}}{\partial \vartheta} , \tag{3.7}$$

$$H^k \equiv H^k(\underset{\sim}{B}, \underset{\sim}{\vartheta}, \underset{\sim}{x}) \equiv \frac{\partial T^{km}}{\partial x^m} .$$

Now on the assumption that the response functions and the body force are continuous, the jump of (3.6) yields immediately the following propagation condition:

$$Q^k_m(\underset{\sim}{n})\, s^m = \rho W^2 s^k \tag{3.8}$$

where $\underset{\sim}{Q} = \underset{\sim}{Q}(\underset{\sim}{n})$ is the acoustic tensor defined by

$$Q^k_m(\underset{\sim}{n}) \equiv 2H^{kp}_{\ \ mf}\, B^{fr} n_p\, n_r , \tag{3.9}$$

In deriving (3.8) and (3.9) we have made use of the compatibility conditions (3.1). It can be shown that from the restrictions due to the principle of universal dissipation (eq. (2.3) of [0]) $\underset{\sim}{Q}(\underset{\sim}{n})$ ia a symmetric tensor.

From (3.9) we see that $\underset{\sim}{Q}(\underset{\sim}{n})$ for each direction $\underset{\sim}{n}$, depends on ($\underset{\sim}{B}$, ϑ, $\underset{\sim}{x}$)but not explicitly on grad $\underset{\sim}{B}$ or grad ϑ . As a result, the inhomogeneites have no direct ef fect on the propagation condition. Also since (3.8) is strictly a local condition, in deriving it we need not use the global field equation established in the preceding paper [0].

Now let $\mathbf{B}$ have the principal form

$$\mathbf{B} = b_1 \mathbf{e}_1 \otimes \mathbf{e}_1 + b_2 \mathbf{e}_2 \otimes \mathbf{e} + b_3 \mathbf{e}_3 \otimes \mathbf{e}_3 , \tag{3.10}$$

where b_i and $\mathbf{e}_i$ generally depend on the position $\mathbf{x}$. Then from the symmetry condition $(2.3)_1$ the stress tensor $\mathbf{T}$ also has the principal form

$$\mathbf{T} = t_1 \mathbf{e}_1 \otimes \mathbf{e}_1 + t_2 \mathbf{e}_2 \otimes \mathbf{e}_2 + t_3 \mathbf{e}_3 \otimes \mathbf{e}_3 , \tag{3.11}$$

where the principal stresses t_i are functions of the principal stretches b_j and the position $\mathbf{x}$. Suppose that the unit normal $\mathbf{n}$ coincides with one of the principal directions $\mathbf{e}_i$, say $\mathbf{n} = \mathbf{e}_1$. Then the propagation condition (3.9) reduces to

$$Q\langle km\rangle = 2H\langle k1m1\rangle b_1 \tag{3.12}$$

where the bracket $\langle\, ,\rangle$ denotes the physical components of the tensors with respect to the principal basis $\{\mathbf{e}_i\}$. For definiteness, let us call these physical components <u>principal components</u> of the tensors also. Then it can be shown that from the symmetry condition $(2.3)_1$ the principal component matrix $[Q\langle km\rangle]$ given by (3.12) is a diagonal matrix, <u>viz</u>,

$$[Q\langle km\rangle] = 2b_1 \,\mathrm{diag}\,[H\langle 1111\rangle, H\langle 1212\rangle, H\langle 1313\rangle]. \tag{3.13}$$

Thus the use of the term principal components does not cause any ambiguity here.

From (3.8) the principal values of $\underset{\sim}{Q}(\underset{\sim}{n})$ are proportional to the squared wave speeds in the direction $\underset{\sim}{n}$, the corresponding principal directions of $\underset{\sim}{Q}(\underset{\sim}{n})$ being the possible directions of the amplitude vectors. In the case $\underset{\sim}{n} = \underset{\sim}{e}_1$, we have shown by (3.13) that $\{\underset{\sim}{e}_i\}$ is a principal basis for $\underset{\sim}{Q}(\underset{\sim}{e}_1)$. Consequently if $Q(\underset{\sim}{n})$ is non-degenerate, then the possible amplitude vectors are either parallel or perpendicular to the normal $\underset{\sim}{n} = \underset{\sim}{e}_1$; the former case corresponds to a longitudinal wave while the latter case corresponds to some transverse waves.

For definitess a wave is called a principal wave if its unit normal is a proper vector of $\underset{\sim}{B}$. What we just observed then amounts to a result of Truesdell: every principal wave is either longitudinal or trasverse. Using Truesdell's notation, we denote the wave speed of a principal wave with $\underset{\sim}{n} = \underset{\sim}{e}_i$ and $\underset{\sim}{a}$ parallel to $\underset{\sim}{e}_j$ by W_{ij}. Then from (3.13) and (3.8) we have

$$W_{ij}^2 = \frac{2}{\rho} H\langle ijij\rangle, \tag{3.14}$$

where i and j are not summed. Hence in order that the wave be possible, $H\langle ijij\rangle$ must be positive.

Differentiating the principal form (3.11), we can express the principal components $H\langle ijij\rangle$ in terms of the principal stretches and their derivatives. The results are

$$H\langle ijij\rangle = \frac{\partial t_i}{\partial b_j}$$

(3.15)
$$H\langle ijij\rangle = \begin{cases} \dfrac{1}{2}\left(\dfrac{t_i - t_j}{b_i - b_j}\right) & \text{if } b_i \neq b_j, \\[2ex] \dfrac{1}{2}\left(\dfrac{\partial t_i}{\partial b_i} - \dfrac{\partial t_i}{\partial b_j}\right) & \text{if } b_i = b_j, \end{cases}$$

where i, j are unequal free indices.

4. Displacement Derivative of the Amplitude Vector

By definition, the displacement derivative of a field on a moving surface is the time derivative of that field along normal trajectories of the surface. To determine the displacement derivative of the amplitude vector with respect to the wave front we differentiate the field equation (3.6) with respect to t, yielding

(4.1)
$$\begin{aligned} &H^{km}{}_{pf}\frac{d}{dt}\left(\frac{\partial B^{pf}}{\partial x^m}\right) + \left(H^{km}{}_{pfij}\dot{B}^{ij} + H^{km}_{\vartheta}{}_{pf}\dot{\vartheta} + \frac{\partial H^{km}{}_{pf}}{\partial x^i}v^i\right)\frac{\partial B^{pf}}{\partial x^m} + \\ &+ H^{km}_{\vartheta}\frac{d}{dt}(g_m) + \left(H^{km}_{\vartheta}{}_{pf}\dot{B}^{pf} + H^{km}_{\vartheta^2}\dot{\vartheta} + \frac{\partial H^{km}}{\partial x^i}v^i\right)g_m + \\ &+ \left(H^{k}{}_{pf}\dot{B}^{pf} + H^{k}_{\vartheta}\dot{\vartheta} + \frac{\partial H^{k}}{\partial x^i}v^i\right) + \dot{\rho}b^k + \rho\dot{b}^k = \dot{\rho}a^k + \rho\dot{a}^k, \end{aligned}$$

where

$$
\begin{aligned}
H^{km}{}_{pf\,ij} &\equiv H^{km}{}_{pf\,ij}(\underset{\sim}{B}, \underset{\sim}{\vartheta}, \underset{\sim}{x}) \equiv \frac{\partial^2 T^{km}}{\partial B^{pf} \partial B^{ij}} , \\
H^{km}_{\vartheta\;pf} &\equiv H^{km}_{\vartheta\;pf}(\underset{\sim}{B}, \underset{\sim}{\vartheta}, \underset{\sim}{x}) \equiv \frac{\partial^2 T^{km}}{\partial B^{pf} \partial \vartheta} , \\
H^{km}_{\vartheta^2} &\equiv H^{km}_{\vartheta^2}(\underset{\sim}{B}, \underset{\sim}{\vartheta}, \underset{\sim}{x}) \equiv \frac{\partial^2 T^{km}}{\partial \vartheta^2} , \\
H^{k}{}_{pf} &\equiv H^{k}{}_{pf}(\underset{\sim}{B}, \underset{\sim}{\vartheta}, \underset{\sim}{x}) \equiv \frac{\partial^2 T^{km}}{\partial x^m \partial B^{pf}} \\
H^{k}_{\vartheta} &\equiv H^{k}_{\vartheta}(\underset{\sim}{B}, \underset{\sim}{\vartheta}, \underset{\sim}{x}) \equiv \frac{\partial^2 T^{km}}{\partial x^m \partial \vartheta}
\end{aligned}
\tag{4.2}
$$

Now on the assumption that the response functions and the time derivative of the body force are continuous as before, the jump of (4.1) yields

$$
\begin{aligned}
& H^{km}{}_{pf}\left[\frac{d}{dt}\left(\frac{\partial B^{pf}}{\partial x^m}\right)\right] + H^{km}{}_{pf\,ij}\left[\dot{B}^{ij}\frac{\partial B^{pf}}{\partial x^m}\right] + H^{km}_{\vartheta}\left[\frac{d}{dt}\left(\frac{\partial \vartheta}{\partial x^m}\right)\right] \\
& + \left(H^{km}_{\vartheta\;pf}\, g_m + H^{k}{}_{pf}\right)\left[\dot{B}^{pf}\right] + \left[\dot{\rho}\right] b^k = \left[\dot{a}^k\right] + \left[\dot{\rho}\, a^k\right] .
\end{aligned}
\tag{4.3}
$$

This jump condition is used to determine the jump $\left[\dfrac{\partial^2 \underset{\sim}{v}}{\partial t^2}\right]$ in the iterated condition of compatibility (3.3) so as to obtain a formula for $\dfrac{\delta \underset{\sim}{s}}{\delta t}$.

To calculate the jumps in (4.3) we use the usual formula for the material derivative, obtaining

$$\rho\left[\dot{a}^{k}\right] = \rho\left[\frac{\partial^{2} v^{k}}{\partial t}\right] - \frac{\rho}{W} s^{k} s^{j} n_{j}$$

$$\left[\frac{d}{dt}\left(\frac{\partial B^{pf}}{\partial x^{m}}\right)\right] = B^{pk}\left[\frac{\partial^{2} v^{f}}{\partial x^{k}\partial x^{m}}\right] + B^{fk}\left[\frac{\partial^{2} v^{p}}{\partial x^{k}\partial x^{m}}\right] - \frac{1}{W} D^{pk}_{m} s^{f} n_{k}$$

$$- \frac{1}{W} D^{fk}_{m} s^{p} n_{k} + \frac{1}{W} D^{pf}_{k} s^{k} n_{m} \frac{2}{W^{3}} B^{kl} s^{p} s^{f} n_{k} n_{m} n_{l}, \tag{4.4}$$

where

$$D^{km}_{p} \equiv \left(\frac{\partial B^{km}}{\partial x^{p}}\right)^{+} \tag{4.5}$$

which characterizes the inhomogeneity of the material in front of the wave. Now using the product rule for jump discontinuities and the continuity equation, we have

$$\left[\dot{\rho} a^{k}\right] = \frac{\rho}{W} s^{k} s^{j} n_{j},$$

$$\left[\dot{B}^{ij} \frac{\partial B^{pf}}{a x^{m}}\right] = - \frac{1}{W}\left(B^{jk} s^{i} + B^{ik} s^{j}\right) D^{pf}_{m} n_{k} \tag{4.6}$$

$$- \frac{1}{W^{3}}\left(B^{ik} s^{j} + B^{jk} s^{i}\right)\left(B^{lf} s^{p} + B^{lp} s^{f}\right) n_{k} n_{l} n_{m}.$$

Since we assume that the region in front of the region is in equilibrium, we have also

$$W\left[\dot{\rho}\right] b^k = -\left(H^{km}{}_{pf} D^{pf}_m + H^{km}_{\vartheta} g_m + H^k\right) s^j n_j . \qquad (4.7)$$

Next we recall that the balance equation for energy is

$$\operatorname{div} \underset{\sim}{f} + \rho r = \rho \vartheta \dot{\eta} . \qquad (4.8)$$

Substituting the constitutive equations (2.3) into this balance equation and then taking the jump, we obtain

$$\frac{\partial f^i}{\partial g_j}\left[\frac{\partial^2 \vartheta}{\partial x^i \partial x^j}\right] + \frac{\partial f^i}{\partial B^{jk}}\left[\frac{\partial B^{jk}}{\partial x^i}\right] = \rho \vartheta \frac{\partial \eta}{\partial B^{ij}}\left[\dot{B}^{ij}\right], \qquad (4.9)$$

where we have assumed that the energy supply r is continuous. Like the displacement field $\underset{\sim}{x}$,the temperature field ϑ obeys also geometric and kinematic conditions of compatibility:

$$\left[\frac{\partial^2 \vartheta}{\partial x^i \partial x^j}\right] = \left[\frac{\partial^2 \vartheta}{\partial x^k \partial x^l}\right] n^k n^l n^i n^j ,$$

$$\left[\frac{\partial^2 \vartheta}{\partial t \partial x^m}\right] = - W\left[\frac{\partial^2 \vartheta}{\partial x^k \partial x^m}\right] n^k . \qquad (4.10)$$

Since we assume that the temperature field is steady in front of the wave, we have also

$$\left[\frac{d}{dt}\left(\frac{\partial \vartheta}{\partial x^m}\right)\right] = \left[\frac{\partial^2 \vartheta}{\partial t \partial x^m}\right] \qquad (4.11)$$

Substituting (4.10)and (4.11) into (4.9), we finally obtain

$$(4.12)\quad \left[\frac{d}{dt}\left(\frac{\partial\vartheta}{\partial x^m}\right)\right] = -\frac{\rho W\vartheta}{k(\underline{n})}\frac{\partial\eta}{\partial B^{ij}}\left[\dot{B}^{ij}\right]n_m - \frac{W}{k(n)}\frac{\partial f^i}{\partial B^{jk}}\left[\frac{\partial B^{jk}}{\partial x^i}\right]n_m,$$

where $k(\underline{n})$ is defined by (1.1). Then we can use equation (3.1) and express the jump $\left[\frac{d}{dt}\left(\frac{\partial\vartheta}{\partial x^m}\right)\right]$ in terms of the amplitude.

Having calculated all the jump terms in equation (4.3), we can now substituting (4.4), (4.6), (4.7), and (4.12) into (4.3) and then substituting the result into (3.3) yielding

$$2\rho\frac{\delta s^k}{\delta t} - \frac{\rho s^k}{W}\frac{\delta W}{\delta t} = \left(Q^k_f(\underline{n}) - \rho W^2\delta^k_f\right)C^f + \frac{2}{Wk_{(\underline{n})}}H^{ki}_{\vartheta}\frac{\partial f^m}{\partial B^{pf}}B^{sp}s^f n_s n_m n_i - \frac{1}{W}H^{km}_{\vartheta}g_{mj}s^j n_j$$

$$-\frac{2}{W}H^{km}_{\vartheta\ pf}B^{fs}g_{mj}s^p n_s - \frac{\vartheta}{k_{(\underline{n})}}H^{km}_{\vartheta}H^{pf}_{\vartheta}s_f n_m n_p$$

$$(4.13)\quad -H^{km}{}_{pf\ell j}\left\{\frac{2}{W}D^{pf}_m B^{\ell r}s^j n_r + \frac{4}{W^3}B^{\ell r}B^{sp}s^j s^f n_m n_r n_s\right\}$$

$$-H^{km}{}_{pf}\left\{\frac{2}{W}D^{pr}_m s^f n_r - \frac{1}{W}D^{pf}_r s^r n_m + \frac{1}{W}D^{pf}_m s^r n_r + \frac{2}{W^3}B^{rs}s^p s^f n_m n_r n_s\right\}$$

$$-2H^{km}{}_{pf}B^{pr}\left\{n_r\frac{\partial x_m}{\partial y_\Gamma}\frac{\partial}{\partial y^\Gamma}\left(\frac{s^f}{W}\right) + n_m\frac{\partial x_r}{\partial y_\Gamma}\frac{\partial}{\partial y^\Gamma}\left(\frac{s^f}{W}\right) - \left(\frac{s^f}{W}\right)\Omega_{\Gamma\Delta}\frac{\partial x_m}{\partial y_\Gamma}\frac{\partial x_p}{\partial y_\Delta}\right\}.$$

This is the general equation of growth for amplitudes of acceleration waves propagating into a steady inhomogeneous region.

The general equation (4.13) simplifies considerably for principal waves. For longitudinal wave with

$$\underset{\sim}{s} = s_1 \underset{\sim}{e}_1 \quad , \quad \underset{\sim}{n} = \underset{\sim}{e}_1, \tag{4.14}$$

we have

$$2\rho \frac{\delta s_1}{\delta t} - \frac{\rho s_1}{W_{11}} \frac{\delta W_{11}}{\delta t} = \frac{2 s_1}{W_{11}} \Big\{ \Big((b_1 + b_2) H\langle 1212\rangle + b\, H\langle 1122\rangle \Big) \Omega\langle 22\rangle +$$

$$+ \Big((b_1 + b_3) H\langle 1313\rangle + b_3 H\langle 1133\rangle \Big) \Omega\langle 33\rangle - H\langle 1111\rangle D\langle 111\rangle -$$

$$- H\langle 111111\rangle D\langle 111\rangle b_1 - 2H\langle 121211\rangle D\langle 122\rangle b_1 - 2H\langle 131311\rangle D\langle 133\rangle b_1 -$$

$$- H\langle 112211\rangle D\langle 221\rangle b_1 - H\langle 113311\rangle D\langle 331\rangle b_1 - H_\vartheta\langle 1111\rangle g\langle 1\rangle b_1 -$$

$$- \frac{1}{2} \frac{\partial t_1}{\partial \vartheta} g\langle 1\rangle - \frac{\vartheta W_{11}}{2k_{(n)}} \left(\frac{\partial t_1}{\partial \vartheta}\right)^2 + \frac{1}{k_{(n)}} \frac{\partial t_1}{\partial \vartheta} J\langle 111\rangle b_1 \Big\} -$$

$$- \frac{2 s_1^2}{W_{11}^3} \Big(H\langle 111\rangle + 2H\langle 111111\rangle b_1 \Big) b_1 \,. \tag{4.15}$$

where

$$J\langle 111\rangle = \frac{\partial f^m}{\partial B^{pf}} n_m n^p n^f \,. \tag{4.16}$$

Similary, for transverse wave with

$$s = s_2 \underset{\sim}{e}_2 \qquad \underset{\sim}{n} = \underset{\sim}{e}_1 \tag{4.17}$$

we have

$$
\begin{aligned}
2\rho\frac{\delta s_2}{\delta t} - \frac{\rho s_2}{W_{12}}\,\frac{\delta W_{12}}{\delta t} = -\frac{2 s_2}{W_{12}}\Big\{&\big(b_1 H\langle 2211\rangle + b_2 H\langle 1212\rangle - b_2 H\langle 2222\rangle\big)\Omega\langle 22\rangle - \\
&- H\langle 2323\rangle b_3 \Omega\langle 33\rangle + H\langle 1212\rangle D\langle 111\rangle + H\langle 121211\rangle D\langle 111\rangle b_1 + \\
&+ H\langle 121222\rangle D\langle 221\rangle b_1 + H\langle 121233\rangle D\langle 331\rangle b_1 + \\
&+ 2H\langle 221212\rangle D\langle 122\rangle b_1 + H_{\vartheta}\langle 1212\rangle g\langle 1\rangle b_1 \Big\}.
\end{aligned}
\tag{4.18}
$$

The equations (4.15) and (4.18) have been solved for waves propagating into some laminated regions, [cf. 1] .

REFERENCES

[1] Ray M. Bowen and C.C. Wang: "Thermodynamical Effects on Acceleration Waves in Isotropic Inhomogeneous Thermoelastic Bodies", Arch. Rational Mech. Anal. **41**, 287-318, 1971.

[2] C. Truesdell and R.A. Toupin: "The Classical Field Theories", Handbuch der Physik, Vol III/1 Springer-Verlag, 1960.

Contents

Part One: Field Equations for Thermoelastic Bodies with Uniform Symmetry

	Page
Preface	3
1. Introduction	5
2. Constitutive Relations of a Thermoelastic Material	7
3. Geometric Structure of a Smooth Thermoelastic Body with Uniform Symmetry	8
4. Field Equations	11
References	20

Part Two: Acceleration Waves in Isotropic Thermoelastic Bodies

Preface	21
1. Introduction	23
2. The Characteristic Riemannian Metric	24
3. Propagation Conditions	27
4. Displacement Derivative of the Amplitude Vector	32
References	39
Contents	41

MIX
Papier aus verantwortungsvollen Quellen
Paper from responsible sources
FSC® C105338

If you have any concerns about our products,
you can contact us on
ProductSafety@springernature.com

In case Publisher is established outside the EU,
the EU authorized representative is:
Springer Nature Customer Service Center GmbH
Europaplatz 3, 69115 Heidelberg, Germany

Printed by Libri Plureos GmbH
in Hamburg, Germany